1 MONTH OF
FREE
READING

at

www.ForgottenBooks.com

By purchasing this book you are eligible for one month membership to ForgottenBooks.com, giving you unlimited access to our entire collection of over 1,000,000 titles via our web site and mobile apps.

To claim your free month visit: www.forgottenbooks.com/free897395

ISBN 978-0-266-84052-7
PIBN 10897395

This book is a reproduction of an important historical work. Forgotten Books uses
state-of-the-art technology to digitally reconstruct the work, preserving the original format
whilst repairing imperfections present in the aged copy. In rare cases, an imperfection in
the original, such as a blemish or missing page, may be replicated in our edition. We do,
however, repair the vast majority of imperfections successfully; any imperfections that
remain are intentionally left to preserve the state of such historical works.

55-84-13 (Revised)

NAVAL POSTGRADUATE SCHOOL
Monterey, California

THREE POSITION ESTIMATION PROCEDURES

by

R. N. FORREST

June 1984

(Revised February 1986)

Approved for public release; distribution unlimited.
Prepared for: Chief of Naval Operations,
Washington, DC 20350

UNCLASSIFIED

SECURITY CLASSIFICATION OF THIS PAGE (When Data Entered)

REPORT DOCUMENTATION PAGE	READ INSTRUCTIONS BEFORE COMPLETING FORM

1. REPORT NUMBER	2. GOVT ACCESSION NO.	3. RECIPIENT'S CATALOG NUMBER
NPS55-84-13 (Revised)		

4. TITLE (and Subtitle)	5. TYPE OF REPORT & PERIOD COVERED
Three Position Estimation Procedures	
	6. PERFORMING ORG. REPORT NUMBER

7. AUTHOR(s)	8. CONTRACT OR GRANT NUMBER(s)
R. N. Forrest	

9. PERFORMING ORGANIZATION NAME AND ADDRESS	10. PROGRAM ELEMENT, PROJECT, TASK AREA & WORK UNIT NUMBERS
Naval Postgraduate School Monterey, CA 93943	TASK 19422 N0003083WR32406

11. CONTROLLING OFFICE NAME AND ADDRESS	12. REPORT DATE
Chief of Naval Operations Washington, D. C. 20350	(Revised February 1986) June 1984
	13. NUMBER OF PAGES. 48

14. MONITORING AGENCY NAME & ADDRESS(If different from Controlling Office)	15. SECURITY CLASS. (of this report)
	Unclassified
	15a. DECLASSIFICATION/DOWNGRADING SCHEDULE

16. DISTRIBUTION STATEMENT (of this Report)

Approved for public release; distribution unlimited.

17. DISTRIBUTION STATEMENT (of the abstract entered in Block 20, if different from Report)

18. SUPPLEMENTARY NOTES

19. KEY WORDS (Continue on reverse side if necessary and identify by block number)

position estimation   composite position estimate
navigational error   composite position fix
bearings only position estimate   line of position fix
bearings only fix   position estimate programs

20. ABSTRACT (Continue on reverse side if necessary and identify by block number)
The report describes three position estimation procedures. The
first is for estimates based on bearings taken on or from
stations with uncertain locations. The second is for use with
two or more lines of position. The third is for combining
estimates from different sources. The initial version of the
report contained an error in a relation relating to the third
procedure. Appendicies to the report contain a description
and listing for a program that can be used to implement the first

DD FORM 1 JAN 73 1473   EDITION OF 1 NOV 65 IS OBSOLETE   UNCLASSIFIED
S/N 0102-LF-014-6601

SECURITY CLASSIFICATION OF THIS PAGE (When Data Entered)

two procedures and for a program to implement the third
procedure.

This revised report contains a correction to the expression for $\sigma_{\hat{X}}^2$ at the bottom of page 14. In addition, it contains a development in an added Appendix 1 of a procedure associated with the position estimation procedures discussed in Sections II and III of the report. It also contains a description and a program listing in an added Appendix 2 for a program called PEST that can be used to implement the first two position estimation procedures and for a program called COMP that can be used to implement the position estimation procedure discussed in Section IV of the report. Both programs are written in BASIC for a Radio Shack TRS-80 Model 100 Portable Computer.

TABLE OF CONTENTS

I.   Introduction

This report describes position estimation procedures that are based on models which relate positional uncertainty to various measurement and estimation errors. The procedures were developed to be used in analyses that relate the effect of positional uncertainty on tactical performance through such factors as weapon accuracy as well as to be used operationally.

In the models, positions are on a plane surface (flat earth model). Because of this condition, the models are not intended for use in situations where the earth's figure is significant. In addition, positional errors are determined by independent normally distributed random variables with known means, variances and covariances. The support for this condition, other than its mathematical convenience, is that it has been used by others, for example, see Reference 1. To use the models, one is required to specify the means, variances and covariances of the error random variables. For these models, this can be done by specifying the systematic errors (biases) and the error magnitudes (standard deviations).

The procedure that is described in Section II relates position estimates that are based on bearings on or from stations to station position uncertainty. One application for the model is the analysis of the effect of sonobuoy position uncertainty on position estimates determined with passive directional sonobuoys.

The procedure that is described in Section III relates position estimates that are based on lines of position to line of position uncertainty. In addition to estimating the uncertainty in position estimates based on celestial observations, the procedure could be used to determine error ellipses for LORAN fixes if the standard deviation values required by the procedure could be obtained.

The procedure that is described in Section IV can be used to combine position estimates from various sources. The procedure which is based on conditions that should not be too restrictive in most cases provides both a composite position estimate and error ellipse.

II. Station Position Uncertainty and Position Estimates

The procedure that is described in this section relates
position estimates that are based on bearings on or from stations
to station position uncertainty. The procedure is based on a
model that is an extension of one that is described in Appendix
1. The model is defined as follows: Each station position error
is determined by an independent bivariate normal distribution
with a zero mean vector and a known covariance matrix. Observed
bearing lines on or from a station are parallel to true bearing
lines. The distance of each observed bearing line from its true
bearing line is determined by an independent normal distribution
with a zero mean and a standard deviation σ.

Because a station's position error is determined by a bivariate normal distribution, the perpendicular distance s between a
line at the assumed location of an observed bearing line and the
observed bearing line is determined by a normal distribution.
The relation between the bivariate normal distribution that
describes the station's position uncertainty is indicated in
Figure 1. In the figure, the positive y-axis direction is north,
the positive x-axis direction is east, and the origin of the
coordinate system is at the assumed station position. The x'y'-
coordinate system is oriented so that the positive x'axis is
coincident with the major axes of the elliptical contours of
the bivariate normal distribution that determines the station
position error and so that the bearing δ of the positive x'-axis

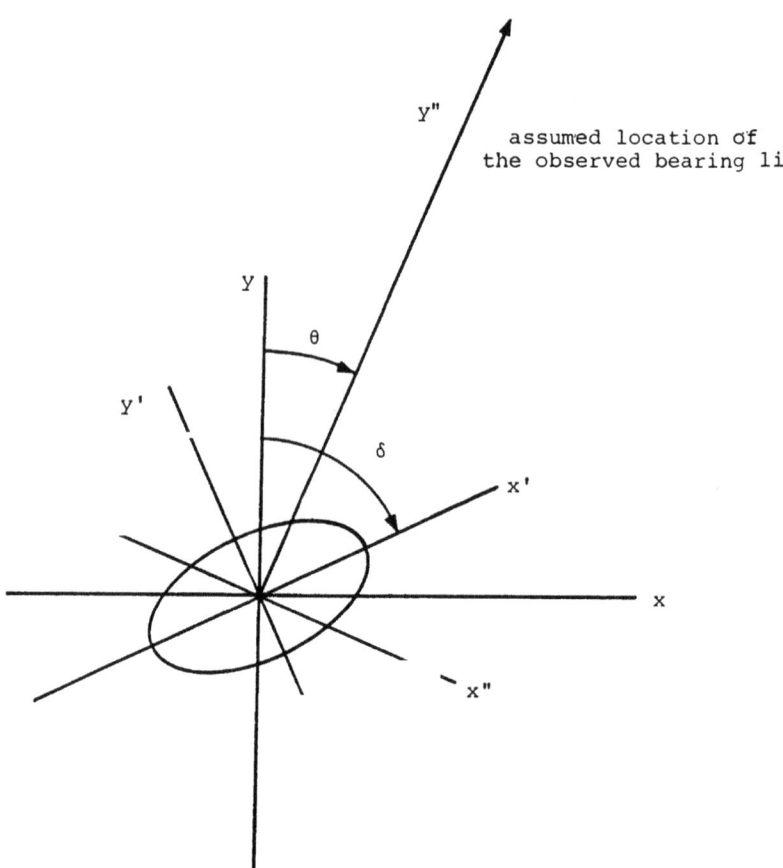

Figure 1. The geometry associated with the determination of σ_s, the standard deviation of the normal distribution that determines the distance between the assumed location of an observed bearing line and the observed bearing line. The assumed (mean) position of the station is at the origin. The ellipse represents a contour on the probability density surface of the bivariate normal distribution that determines the station's position.

4

satisfies the condition: $0° \leq \delta < 180°$. The angle θ is an observed bearing. The x"y"-coordinate system is oriented so that the positive y"-axis is in the direction of the observed bearing line. As a consequence of these relationships,

$$\sigma_s{}^2 = \sigma_{x'}{}^2 \sin^2(\delta-\theta) + \sigma_{y'}{}^2 \cos^2(\delta-\theta)$$

where $\sigma_{x'}{}^2$ and $\sigma_{y'}{}^2$ are the elements of the station position error covariance matrix relative to the x'y'-coordinate system.

A procedure for determining position estimates that are based on bearings on or from stations at known positions is described in Appendix 1 of this report. The procedure is based on a model that is equivalent to one that is described in Reference 2. The procedure that is described in this section is based on a model that is an extension of it.

The procedure in Appendix 1 of this report is based on a model in which a station's bearing error is determined by a normal distribution with zero mean (bias) and standard deviation e. The bearing error is related to the distance on a circular arc between a station's true bearing line and observed bearing line. The arc is on the circle with its center at the station that passes through an initial estimate of an object's position. This distance is determined by a normal distribution with mean zero and standard deviation $\sigma = re$ where r is the range of the initial estimate from the station and the standard deviation e is measured in radians. In the model, arc distance is approximated using a first order approximation which in effect replaces the circle with its tangent line at the initial estimate. As a consequence, the distance u on the tangent line between the observed bearing line and the true bearing line is determined by a normal random variable with mean zero and standard deviation σ. The distance u can be expressed in terms of w, the distance on the tangent line between the observed bearing line and the initial estimate that is also determined by a normal random variable with standard deviation σ, and v, the distance on the tangent line between the true bearing line and the initial estimate. And, as shown in Appendix 1 of this report, this distance can be expressed in terms of the unknown coordinates of the object's position.

The effect of station position uncertainty is accounted for by the distance s between the observed bearing line and the assumed observed bearing line. In the model s is determined by a normal distribution with mean zero and standard deviation σ_s as given above. This approximation is consistent with the first order approximation of arc distance. As a consequence of these two approximations, all of the bearing lines are replaced by lines parallel to the line joining the initial estimate's position and the station's assumed position, both of which are known positions. The geometry involved in shown in Figure 2. The modified relationships resulting from the introduction of station position uncertainty are shown in Figure 3. The

6

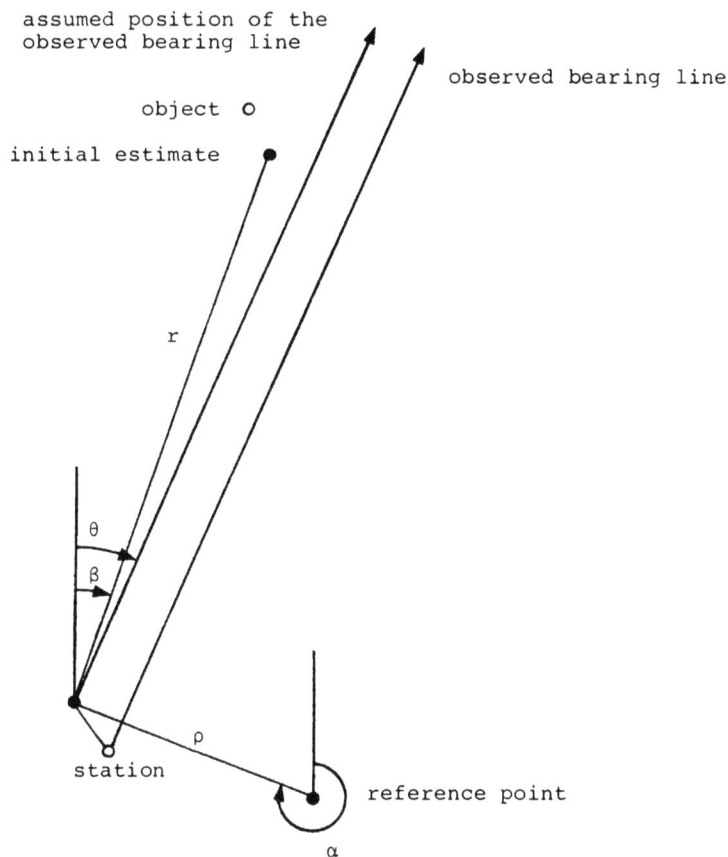

Figure 2. The geometry of the position estimation model. The bearing β and range r of the initial estimate from the assumed station position have the role of β and r in Appendix 1 of this report.

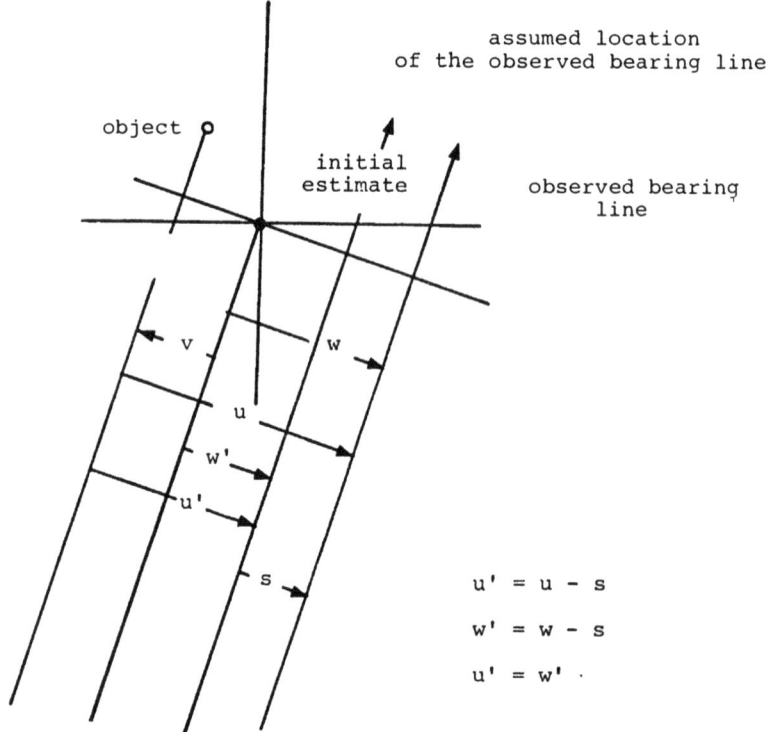

Figure 3. The quantities u, v and w correspond to the quantities
u, v and w in Appendix 1 of Reference 3. The auxiliary
quantities u' and w' are defined in the figure. The
replacement of σ^2 by $\sigma^2 + \sigma_s^2$ in the procedure in
Appendix 1 of this report is justified by noting that
with this substitution u' and w' are equivalent to u
and w.

important thing to note is that u' = u - s is determined by a normal distribution with mean zero and standard deviation $[\sigma^2 + \sigma_s^2]^{\frac{1}{2}}$ and that otherwise U' is equivalent to U with respect to the procedure in Appendix 1 of this report. As a consequence of this, the procedure can be extended to include station uncertainty by replacing σ by $[\sigma^2 + \sigma_s^2]^{\frac{1}{2}}$ where ever it is used. In this case, $\sigma = re$ where r is the range of an initial estimate of an object's position from the assumed station position and e is the bearing error (standard deviation) in radians of the bearings associated with the station.

III. Line of Position Uncertainty and Position Estimates

The procedure that is described in this section relates position estimates that are based on lines of position to line of position uncertainty. The procedure is based on a model that is defined as follows: Lines of position are straight lines. Observed lines of position are parallel to true lines of position. The distance of an observed line of position from a true line of position is determined by an independent normally distributed random variable with known mean and standard deviation. Lines of position are specified in terms of a rectangular coordinate system with the origin at a reference point as shown in Figure 4. For celestial navigation, an appropriate choice for the reference point would be the assumed position. Since bearing lines are lines of position, this model differs from the model that is described in Section II only in terminology. However, operationally the use of the model that is described in this section differs in the way the standard deviation of the distance of the line from a true line is determined. The standard deviation associated with each line of position must be specified. If this can be done, the procedure can be used. As an example, suppose that the values are σ_1 for the first line of position and σ_2 for the second where $\sigma_1 > \sigma_2$ and that the lines of position intersect at a 90° angle. In this example, the minimum area confidence (probability) region is an ellipse that is centered on the estimated position. And the

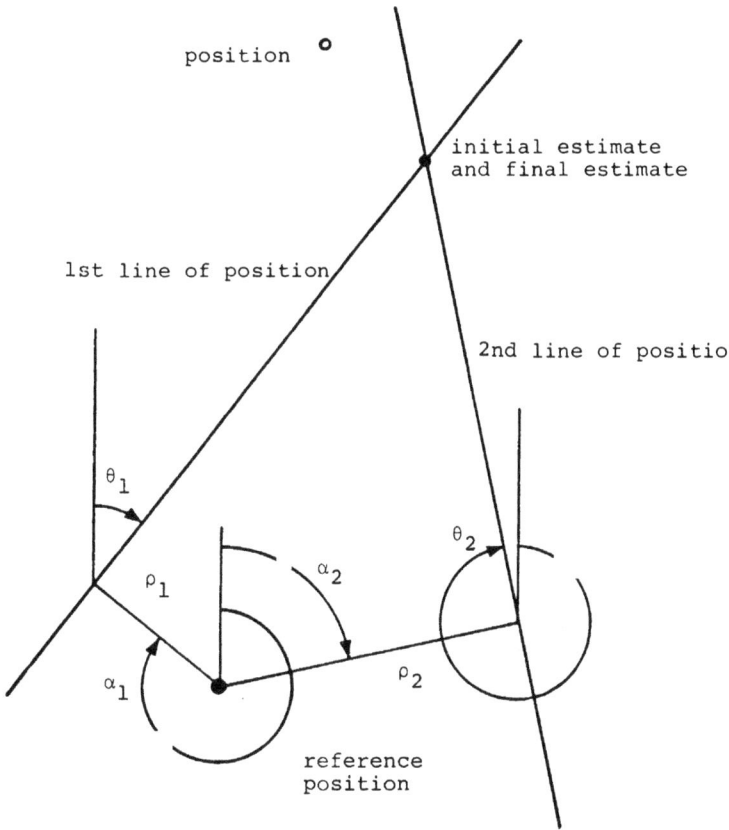

Figure 4. The geometry when only two lines of position are used
with the procedure in Appendix 1 of this report. The lines of
position correspond to bearing lines for observed bearings θ_1 and
θ_2 with respect to the procedure. The procedure determines the
size and the orientation of elliptical confidence regions of confidence p, the location of the initial estimate and the location
of the final estimate. The locations of the estimates correspond
when only two lines are used. If the lines were obtained from
sextant observations, the assumed position should be the reference
position in which case the lines of position would be determined
by the azimuth anges α_1 and α_2 and the distances ρ_1 and ρ_2.

11

estimated position is at the intersection of the bearing lines. For a confidence (probability) of containment of $1 - \exp(-k^2/2)$, the major axis of the ellipse is coincident with the first bearing line and it is of length $2k\sigma_1$, the minor axis of the ellipse is coincident with the second bearing line and it is of length $2k\sigma_2$ and the area of the ellipse is $\pi k^2 \sigma_1 \sigma_2$.

IV. A Composite Position Estimate

The procedure that is described in this section is for
combining position estimates for an object from independent
sources. It is based on the following model: The rectangular
coordinates of each position estimate are determined by an independent bivariate normal distribution whose covariance matrix
is known but whose mean vector is not known. The components of
the mean vector for each of the distributions are the unknown
coordinates x and y of the object. This model implies that the
natural logarithm of the likelihood function for a set of n
estimates can be expressed as follows:

$$\log L = K - 1/2 \sum_{i=1}^{n} (\hat{\underline{x}}_i - \underline{x})' \Sigma_i^{-1} (\hat{\underline{x}}_i - \underline{x})$$

where K is a constant, $\hat{\underline{x}}_i$ is an estimate vector with components $\hat{x}_i$
and $\hat{y}_i$, $\underline{x}$ is the common mean vector with components x and y,
the unknown coordinates of the object, and Σ_i is the covariance
matrix with elements $\sigma_{i\hat{x}}^2$, $\sigma_{i\hat{y}}^2$ and $\sigma_{i\hat{x}\hat{y}}$. The maximum likelihood estimates $\hat{x}$ and $\hat{y}$ of the unknown coordinates x and y are
the solutions of the two simultaneous linear equations determined by

$$\frac{\partial (\log L)}{\partial x} \bigg|_{\substack{x=\hat{x} \\ y=\hat{y}}} = 0 \quad \text{and} \quad \frac{\partial (\log L)}{\partial y} \bigg|_{\substack{x=\hat{x} \\ y=\hat{y}}} = 0 \quad .$$

13

The equations can be written as

$$A \hat{x} + B \hat{y} = D$$
$$B \hat{x} + C \hat{y} = E$$

and their solution as

$$\hat{x} = \frac{CD - BE}{AC - B^2} \qquad \hat{y} = \frac{AE - BD}{AC - B^2}$$

where $A = \Sigma a_i$, $B = \Sigma b_i$, $C = \Sigma c_i$, $D = \Sigma(a_i \hat{x}_i + b_i \hat{y}_i)$,

$E = \Sigma(b_i \hat{x}_i + c_i \hat{y}_i)$, $a_i = \sigma_{i\hat{y}}^2/d_i$, $b_i = -\sigma_{i\hat{x}\hat{y}}/d_i$, $c_i = \sigma_{i\hat{x}}^2/d_i$,

$d_i = \sigma_{i\hat{x}}^2 \sigma_{i\hat{y}}^2 - \sigma_{i\hat{x}\hat{y}}^2$ and all of the sums are for i from 1 to n.

Since they are linear combinations of the estimates $\hat{x}_i$ and $\hat{y}_i$, the estimates $\hat{x}$ and $\hat{y}$ are determined by a bivariate normal distribution. Consequently, all that is required to determine this distribution is its mean vector with components $\mu_{\hat{x}}$ and $\mu_{\hat{y}}$ and its covariance matrix with elements $\sigma_{\hat{x}}^2$, $\sigma_{\hat{y}}^2$ and $\sigma_{\hat{x}\hat{y}}$. The mean vector is determined by

$$\mu_{\hat{x}} = E\{(CD - BE)/(AC - B^2)\} = x$$

and

$$\mu_{\hat{y}} = E\{(AE - BD)/(AC - B^2)\} = y$$

And, the covariance matrix is determined by

$$\sigma_{\hat{x}}^2 = E\{(CD - BE) - E(CD - BE)\}^2/(AC - B^2)^2$$
$$= \{C^2(F+I+2L) - 2CB(H+K+M) + B^2(G+J+2N)\}/(AC - B^2)^2$$

14

$$\sigma_{\hat{y}}^2 = E\{(AE - BD) - E(AE - BD)\}^2/(AC - B^2)^2$$

$$= \{A^2(G+J+2N) - 2AB(H+K+M) + B^2(F+I+2L)\}/(AC - B^2)^2$$

$$\sigma_{\hat{x}\hat{y}} = E\{[(CD-BE)-E(CD-BE)]\cdot[(AE-BD)-E(AE-BD)]\}/(AC - B^2)^2$$

$$= \{(AC+B^2)(H+K+M)-CB(F+I+2L)-BA(G+J+2N)\}/(AC - B^2)^2$$

where $F = \Sigma a_i^2 \sigma_{i\hat{x}}^2$, $G = \Sigma b_i^2 \sigma_{i\hat{x}}^2$, $H = \Sigma a_i b_i \sigma_{i\hat{x}}^2$, $I = \Sigma b_i^2 \sigma_{i\hat{y}}^2$,

$J = \Sigma c_i^2 \sigma_{i\hat{y}}^2$, $K = \Sigma b_i c_i \sigma_{i\hat{y}}^2$, $L = \Sigma a_i b_i \sigma_{i\hat{x}\hat{y}}$, $M = \Sigma (a_i c_i + b_i^2) \sigma_{i\hat{x}\hat{y}}$

and $N = \Sigma b_i c_i \sigma_{i\hat{x}\hat{y}}$ where all of the sums are for i = 1 to n.

By using arguments given in Appendix 1 of this report, one can show that the axes of the elliptical confidence regions associated with $\hat{x}$ and $\hat{y}$ are coincident with an x'y'-coordinate system where the transformation from the xy-coordinate system to this system is the coordinate axes rotation through the angle γ defined by $\tan 2\gamma = 2\sigma_{\hat{x}\hat{y}}/(\sigma_{\hat{y}}^2 - \sigma_{\hat{x}}^2)$. For a confidence p, from Reference 3, the minimum area confidence region is an ellipse with semi-axes $k\,\sigma_{\hat{x}'}$ and $k\,\sigma_{\hat{y}'}$, and area $\pi k^2 \sigma_{\hat{x}'} \sigma_{\hat{y}'}$, where $k = [-2 \ln(1-p)]^{\frac{1}{2}}$,

$$\sigma_{\hat{x}'}^2 = \sigma_{\hat{x}}^2 \cos^2\gamma - 2\sigma_{\hat{x}\hat{y}} \cos\gamma \sin\gamma + \sigma_{\hat{y}}^2 \sin^2\gamma,$$

$$\sigma_{\hat{y}'}^2 = \sigma_{\hat{x}}^2 \sin^2\gamma + 2\sigma_{\hat{x}\hat{y}} \cos\gamma \sin\gamma + \sigma_{\hat{y}}^2 \cos^2\gamma$$

and the center of the ellipse is at the point $(\hat{x},\hat{y})$. In this coordinate system, $\sigma_{\hat{x}'\hat{y}'} = 0$.

The above equations can be used to specify a composite position estimate in terms of the location, orientation and size of an elliptical confidence region which is generally the form in which position estimates of the kind that are being considered here are specified. But, since values of $\sigma_{\hat{x}}^2$, $\sigma_{\hat{y}}^2$ and $\sigma_{\hat{x}\hat{y}}$ are required for each of the n estimates that is being combined, a way is needed for determining these values given the orientation and size of an elliptical confidence region. A procedure to do this when the orientation is given in terms of the direction δ of the major axis and the size is given in terms of the lengths SMJ and SMI of the semi-major and semi-minor axes and the confidence p is described next.

By using an xy-coordinate system in which the positive y-axis direction is north and the positive x-axis direction is east and with the convention $0° \leq \delta < 180°$, the dependence of the value of the rotation angle γ and of the order relation between $\sigma_{\hat{x}'}$ and $\sigma_{\hat{y}'}$ on the value of the major axis direction δ is indicated by the following table:

$0° \leq \delta < 45°$: $\gamma = \delta$ and $\sigma_{\hat{y}'} > \sigma_{\hat{x}'}$

$45° \leq \delta < 135°$: $\gamma = \delta - 90°$ and $\sigma_{\hat{x}'} > \sigma_{\hat{y}'}$

$135° \leq \delta < 180°$: $\gamma = \delta - 180°$ and $\sigma_{\hat{y}'} > \sigma_{\hat{x}'}$

With an order relation and a value for p, values for $\sigma_{\hat{x}'}$ and $\sigma_{\hat{y}'}$ can be determined with values for SMJ and SMI. With values for $\sigma_{\hat{x}'}$, $\sigma_{\hat{y}'}$ and γ, values for $\sigma_{\hat{x}}^2$, $\sigma_{\hat{y}}^2$ and $\sigma_{\hat{x}\hat{y}}$ can be found from the following equations:

$$\sigma_{\hat{x}}^2 = \sigma_{\hat{x}'}^2 \cos^2 \gamma + \sigma_{\hat{y}'}^2 \sin^2 \gamma,$$

$$\sigma_{\hat{y}}^2 = \sigma_{\hat{x}'}^2 \sin^2 \gamma + \sigma_{\hat{y}'}^2 \cos^2 \gamma$$

and

$$\sigma_{\hat{x}\hat{y}} = (\sigma_{\hat{y}'}^2 - \sigma_{\hat{x}'}^2) \sin \gamma \cos \gamma$$

which can be obtained by inverting the equations above for $\sigma_{\hat{x}'}^2$ and $\sigma_{\hat{y}'}^2$.

As an example, suppose the data in the following table represent three independent position estimates:

	$\hat{x}$	$\hat{y}$	δ	SMJ	SMI	k
1st	-3.7	18.1	59°	36	20	2
2nd	11.8	8.4	105°	37	11	2
3rd	0	0	146°	45	23	2

Here, distances are in nautical miles and p = .86 in each case. For this example with values in square nautical miles:

$$\sigma_{1\hat{x}}^2 = 264.58, \quad \sigma_{1\hat{y}}^2 = 159.42 \text{ and } \sigma_{1\hat{x}\hat{y}} = 98.89$$

$$\sigma_{2\hat{y}}^2 = 321.35, \quad \sigma_{2\hat{y}}^2 = 51.15 \text{ and } \sigma_{2\hat{x}\hat{y}} = -78$$

$$\sigma_{3\hat{x}}^2 = 249.20, \quad \sigma_{3\hat{y}}^2 = 389.30 \text{ and } \sigma_{3\hat{x}\hat{y}} = -173.4$$

These values give the following composite estimate:
$\hat{x} = -2.46$ nautical miles and $\hat{y} = 12.26$ nautical miles.

For this case, $\sigma_{\hat{x}}$ = 10.28 nautical miles, $\sigma_{\hat{y}}$ = 6.21 nautical miles and $\sigma_{\hat{x}\hat{y}}$ = -55.1 square nautical miles. And, for k = 2: SMJ = 20.56 nautical miles, SMI = 12.42 nautical miles and δ = 119°. The composite confidence region and its three component confidence regions are shown in Figure 5.

As a second example, suppose each position estimate is determined by a circular normal distribution. Then $\sigma_{i\hat{x}}$ = $\sigma_{i\hat{y}}$ = σ_i and $\sigma_{i\hat{x}\hat{y}}$ = 0 for i = 1 to n. In this case, the composite estimate is:

$$\hat{x} = (\Sigma\hat{x}_i/\sigma_i)/(\Sigma 1/\sigma_i) \ , \ \hat{y} = (\Sigma\hat{y}_i/\sigma_i)/(\Sigma 1/\sigma_i) \ ,$$

$$\sigma_{\hat{x}}^2 = n/(\Sigma 1/\sigma_i)^2 \ , \ \sigma_{\hat{y}}^2 = n/(\Sigma 1/\sigma_i)^2 \text{ and } \sigma_{\hat{x}\hat{y}} = 0.$$

In this example, since $\hat{x}$ and $\hat{y}$ are determined by a circular normal distribution, the minimum area confidence regions are circles and orientation is not an issue.

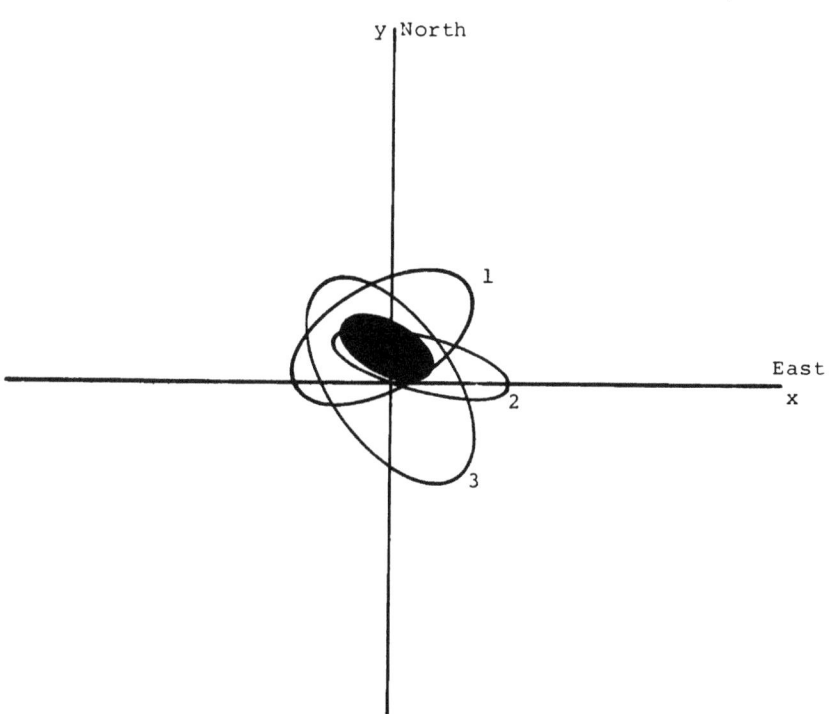

Figure 5. The ellipses define the confidence regions of
the first example. The composite confidence region is in
black. The position estimates are at the center of the
ellipses. The numbers indicate the order of the estimate
in the table on Page 17.

20

APPENDIX 1:   A Position Estimation Procedure

Consider a target whose position is unknown and a set of reference stations whose positions are known. Assume conditions are such that observed target bearings from the stations can be considered to be values of independent normal random variables whose means are equal to the true target bearings and unknown but whose standard deviations are known. The position estimation procedure that is described here is a maximum likelihood estimation procedure that is based on this assumption and the assumption that the conditions are such that the stations and the target can be considered to be located on a plane tangent to the earth's surface at a point in their neighborhood.

With the above assumptions, the likelihood of observed bearings θ_1, θ_2, ..., θ_n from stations labeled 1, 2, ..., n is:

$$L(\theta_1, \theta_2, ..., \theta_n) = \prod_1^n \frac{1}{\sqrt{2\pi}\, e_i} \quad \exp - \frac{1}{2} \sum_1^n (\theta_i - \phi_i)^2 / e_i^2$$

where ϕ_1, ϕ_2, ..., ϕ_n are the unknown station true bearings and e_1, e_2, ..., e_n are the known stations standard deviations.

To a first order approximation, the set of bearing estimates $\hat{\phi}_1$, $\hat{\phi}_2$, ..., $\hat{\phi}_n$ that are determined by the procedure make $L(\theta_1, \theta_2, ..., \theta_n)$ a maximum subject to the constraint that the bearing lines determined by a set of bearing estimates must all pass through a common point. The common point is the estimate of the target's position.

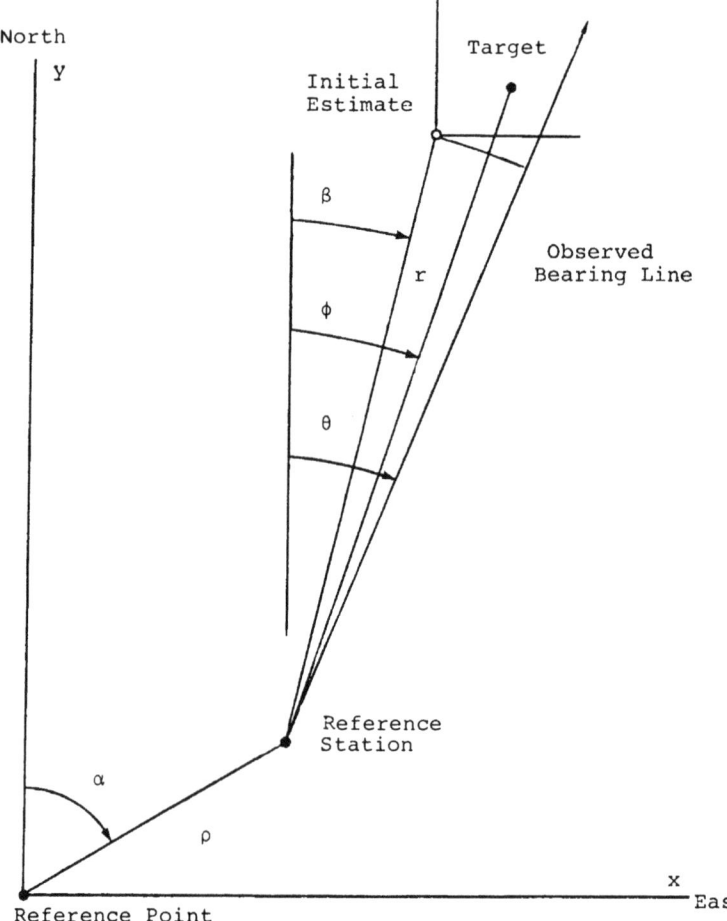

Figure 6. The coordinate geometry. The coordinates of the initial
estimate are (x*,y*). In the development, the reference
point is at the initial estimate.

In order to impose the constraint on the estimate ϕ_1 through ϕ_n, consider the quantities $u_i = r_i(\theta_i - \phi_i)$, $v_i = r_i(\phi_i - \beta_i)$ and $w_i = r_i(\theta_i - \beta_i)$ where i represents any station number from 1 to n. Here, ϕ_i is the target's true bearing, β_i is the bearing of an initial estimate of the target's position, θ_i is an observed target bearing and r_i is the range of the initial estimate from the station. It is also the radius of a circle that is centered on the station and passes through the initial estimate as shown in Figure 6 and u_i, v_i and w_i are related arc lengths on this circle with $u_i = w_i - v_i$. In this relation, w_i is known and v_i can be expressed in terms of the unknown coordinates of the target with an approximation that does not involve ϕ_i. To do this, consider a rectangular coordinate system whose origin is at the position of the initial estimate and whose axes are oriented like those shown in Figure 6. To first order in this system, $v_i = x\cos\beta_i - y\sin\beta_i$ where x and y are the unknown target coordinates and $\theta_i - \phi_i = (\theta_i - \beta_i) - (x\cos\beta_i - y\sin\beta_i)/r_i$. The use of this relation implies that the bearing line determined by β_i, the bearing of the initial estimate, is approximately parallel to the true bearing line determined by ϕ_i. The use of this relation for all stations imposes the constraint on the maximum likelihood bearing estimates by replacing what would otherwise have been estimates of n independent bearings ϕ_1 through ϕ_n by estimates of two independent quantities, the rectangular coordinates x and y. It also implies that the initial estimate's range from a station is approximately the target's range from a station, that is, that the initial estimate's position is relatively close to the target's position.

23

Since $u_i = r_i(\theta_i - \phi_i)$, the likelihood of observed bearings $\theta_1, \theta_2, \ldots, \theta_n$ can be written as:

$$L(\theta_1, \theta_2, \ldots, \theta_n) = \prod_1^n \frac{1}{\sqrt{2\pi}\,\sigma_i} \quad \exp -\frac{1}{2}\sum_1^n u_i^2/\sigma_i^2$$

where $\sigma_i = r_i e_i$ and e_i is the standard deviation of θ_i. The maximum likelihood estimates for ϕ_1 through ϕ_n are determined by the estimates for x and y that make $L(\theta_1, \theta_2, \ldots, \theta_n)$ a maximum. In this case, making $L(\theta_1, \theta_2, \ldots, \theta_n)$ a maximum is equivalent to making $\sum_1^n (u_i^2/\sigma_i^2)$ a minimum. So, to find the maximum likelihood estimates $\hat{x}$ and $\hat{y}$, solve the following two equations for $\hat{x}$ and $\hat{y}$:

$$\frac{\partial(\ln L)}{\partial x}\Big|_{\substack{x=\hat{x}\\y=\hat{y}}} = 0 \quad \text{and} \quad \frac{\partial(\ln L)}{\partial y}\Big|_{\substack{x=\hat{x}\\y=\hat{y}}} = 0$$

With the constraint given by $u_i = w_i - x\cos\beta_i + y\sin\beta_i$ where $w_i = r_i(\theta_i - \beta_i)$, the equations can be written as follows:

$$\sum_1^n (w_i - \hat{x}\cos\beta_i + \hat{y}\sin\beta_i)(\cos\beta_i)/\sigma_i^2 = 0$$

and

$$\sum_1^n (w_i - \hat{x}\cos\beta_i + \hat{y}\sin\beta_i)(\sin\beta_i)/\sigma_i^2 = 0.$$

24

In terms of the following quantities:

$$A = \Sigma(\cos^2 \beta_i)/\sigma_i^2 , \qquad\qquad B = \Sigma(\sin \beta_i \cos \beta_i)/\sigma_i^2 ,$$

$$C = \Sigma(\sin^2 \beta_i)/\sigma_i^2 , \qquad\qquad D = \Sigma(w_i \cos \beta_i)/\sigma_i^2 ,$$

$$E = \Sigma(w_i \sin \beta_i)/\sigma_i^2 ,$$

the equations become:

$$A\hat{x} - B\hat{y} = D$$

$$B\hat{x} - C\hat{y} = E$$

The solutions are:

(1) $$\hat{x} = (BE - CD)/(B^2 - AC)$$

and

(2) $$\hat{y} = (AE - BD)/(B^2 - AC)$$

A confidence region can be constructed about an estimated position. In order to indicate how this can be done, a probability region about the true position will be considered first.

Both $\hat{x}$ and $\hat{y}$ are values of random variables. If a new set of bearings θ_1, θ_2, ..., θ_n is observed (for the same initial estimate and a fixed target), in general, a new pair of values $\hat{x}$ and $\hat{y}$ will be obtained.

If $\hat{X}$ and $\hat{Y}$ represent these random variables, from (1) and (2),

(3) $\hat{X} = \dfrac{1}{(B^2-AC)} \sum\limits_1^n (W_i/\sigma_i^2)(B \sin \beta_i - C \cos \beta_i)$

(4) $\hat{Y} = \dfrac{1}{(B^2-AC)} \sum\limits_1^n (W_i/\sigma_i^2)(A \sin \beta_i - B \cos \beta_i)$

where $W_i = r_i(\Theta_i - \beta_i)$.

Since $\hat{X}$ and $\hat{Y}$ are a linear combination of the n normal random variables W_1, W_2, ..., W_n, or equivalently of the n normal random variables Θ_1, Θ_2, ..., Θ_n, they have a joint normal distribution. Since $E(W_i) = r_i(\phi_i - \beta_i)$, if $\beta_i = \phi_i$ for i = 1, 2, ..., n, that is, if the initial estimate of the target's position is at the target's position, $E(W_1) = 0$ for i = 1, 2, ..., n. In this case $E(\hat{X}) = 0$ and $E(\hat{Y}) = 0$ and the joint normal distribution is centered on the object's position. To the degree of the approximations that have been made, this is also true if the initial estimate is not at the target's position.

A region of minimum area for a given probability of containment of an estimated position can be determined. The region is bounded by an ellipse which is centered on the object's position and whose axes lie along the axes of an x'y'-coordinate system that is obtained by rotating the xy-coordinate system that is centered on the object's position through an angle γ. In this system, $\sigma_{\hat{x}',\hat{y}'}$ is 0, that is $\hat{X}'$ and $\hat{Y}'$ are independent normal random variables. The two coordinate systems are illustrated in

26

Figure 7. The coordinates of a point in the two systems are related by

$$x' = x \cos \gamma - y \sin \gamma$$
$$y' = x \sin \gamma + y \cos \gamma$$

These relations imply:

(5) $\quad \sigma_{\hat{x}'}^2 = \sigma_{\hat{x}}^2 \cos^2 \gamma - 2\sigma_{\hat{x}\hat{y}} \cos \gamma \sin \gamma + \sigma_{\hat{y}}^2 \sin^2 \gamma$,

(6) $\quad \sigma_{\hat{y}'}^2 = \sigma_{\hat{x}}^2 \sin^2 \gamma + 2\sigma_{\hat{x}\hat{y}} \cos \gamma \sin \gamma + \sigma_{\hat{y}}^2 \cos^2 \gamma$

and

(7) $\quad \sigma_{\hat{x}'\hat{y}'} = (\sigma_{\hat{x}}^2 - \sigma_{\hat{y}}^2) \sin \gamma \cos \gamma + \sigma_{\hat{x}\hat{y}} (\cos^2 \gamma - \sin^2 \gamma)$

where γ, the angle of rotation of the coordinate axes, is positive in the clockwise direction. And $\sigma_{\hat{x}'\hat{y}'} = 0$ implies

$$\tan 2\gamma = \frac{2\sigma_{\hat{x}\hat{y}}}{\sigma_{\hat{y}}^2 - \sigma_{\hat{x}}^2}$$

With the initial estimate of the target's position at the target's position $E(W_i) = 0$ and therefore $\mathrm{Var}(W_i) = \sigma_i^2$ for $i = 1, 2, \ldots, n$. In this case, from (3) and (4)

$$\sigma_{\hat{x}}^2 = \frac{1}{(B^2 - AC)^2} \sum_1^n (1/\sigma_i^2)(B \sin \beta_i - C \cos \beta_i)^2,$$

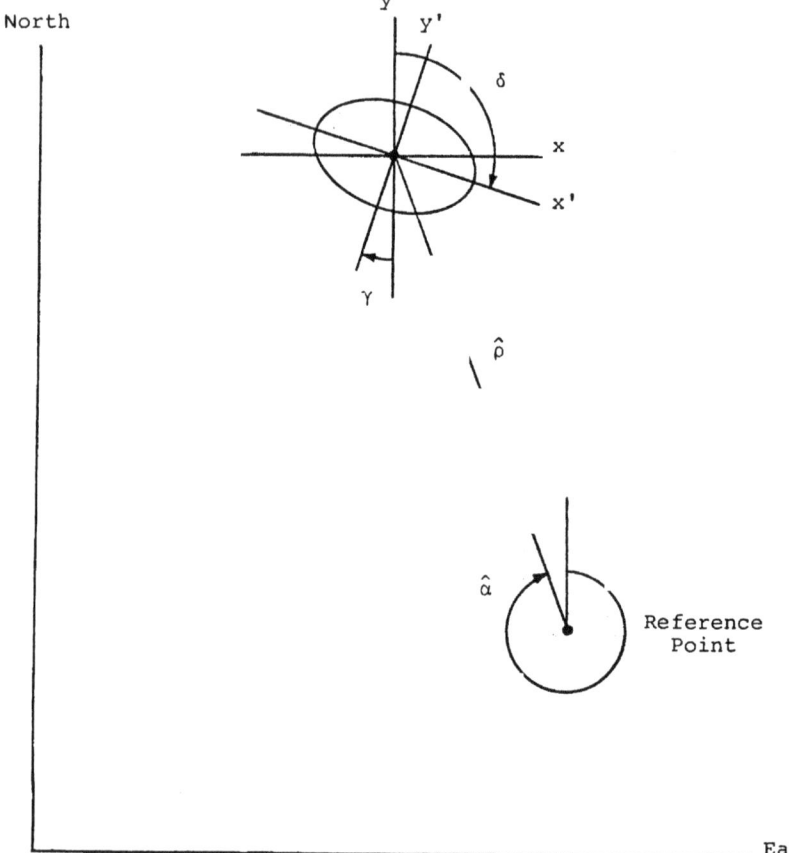

Figure 7. An elliptical confidence region and the primed coordinate system in which the covariance $\sigma_{\hat{x}'\hat{y}'}$ is zero. The center of the ellipse and the origin of the coordinate systems are at the target's estimated position. The estimated bearing $\hat{\alpha}$ and estimated range $\hat{\rho}$ are indicated for a reference point. The major axis direction is δ.

$$\sigma_{\hat{y}}^2 = \frac{1}{(B^2 - AC)^2} \sum_1^n (1/\sigma_i^2)(A \sin \beta_i - B \cos \beta_i)^2$$

and

$$\sigma_{\hat{x}\hat{y}} = \frac{1}{(B^2 - AC)^2} \sum_1^n (1/\sigma_i^2)(B \sin \beta_i - C \cos \beta_i)(A \sin \beta_i - B \cos \beta_i).$$

Using the definition for A, B and C, the above become

(8)
$$\sigma_{\hat{x}}^2 = \frac{C}{(AC - B^2)} \ ,$$

(9)
$$\sigma_{\hat{y}}^2 = \frac{A}{(AC - B^2)} \ ,$$

and

(10)
$$\sigma_{\hat{x}\hat{y}} = \frac{B}{(AC - B^2)} \ .$$

So $\tan 2Y = 2B/(A-C)$ for $\beta_i = \phi_i$, $i = 1, 2, \ldots, n$.

With the target's position known and, consequently, ϕ_i known for $i = 1, 2, \ldots, n$, the above expressions for $\sigma_{\hat{x}}^2$, $\sigma_{\hat{y}}^2$, $\sigma_{\hat{x}\hat{y}}$ and Y can be used, since the initial estimate of the target's position can be taken as the target's position.

With values for $\sigma_{\hat{x}}$, $\sigma_{\hat{y}}$, $\sigma_{\hat{x}\hat{y}}$ and Y, values for $\sigma_{\hat{x}'}$ and $\sigma_{\hat{y}'}$ can be found by using equations (5) and (6). The probability that an estimated position will be within an ellipse of semiaxes $k\sigma_{\hat{x}'}$ and $k\sigma_{\hat{y}'}$ which is centered on the target's position is

1 -exp$(-k^2/2)$. This result can be found by integrating the bivariate normal density over the ellipse. And the area of the
ellipse is $k^2\sigma_{\hat{x}'}\sigma_{\hat{y}'}$.

Given estimates $\hat{x}$ and $\hat{y}$ found by using Equations (1) and (2),
the ellipse with semi-axes $k\sigma_{\hat{x}'}$ and $k\sigma_{\hat{y}'}$ in a coordinate system
that is centered on the point $(\hat{x},\hat{y})$ and has been rotated through
an angle γ is a $1 - \exp(-k^2/2)$ confidence region. This follows,
since, to the degree of the approximations involved, the bivariate normal distribution of X and Y is centered on the target's
position. The confidence ellipse is defined if σ_x^2, σ_y^2 and σ_{xy}
can be found, that is if the elements of the covariance matrix
can be found. To the degree of the approximations involved,
this can be done as follows: First, assume the initial estimate
of the target's position is at the target's position. Then,
values for $\sigma_{\hat{x}}^2$, $\sigma_{\hat{y}}^2$, $\sigma_{\hat{x}\hat{y}}$ and γ can be determined by using Equations
(8), (9) and (10) These values can then be used to determine
$\sigma_{\hat{x}'}^2$, $\sigma_{\hat{y}'}^2$ and $\sigma_{\hat{x}'\hat{y}'}$ by using Equations (5), (6) and (7). Now,
with a value for k, a confidence region can be constructed. To
the degree of the approximations involved, the shape of the confidence region is independent of both the target's position and
of the initial estimate of the target's position.

For the case where bearings are taken from the target on two
or more stations, θ_i is the reciprocal of the bearing taken from
the target.

A discussion of the theory of bearings only position
estimation procedures for situations similar to the one considered
here is given in Reference 2.

The following equations are evaluated in the program to determine the coordinates x* and y* of the initial estimate:

$$x^* \sin (\theta_2-\theta_1) = [\rho_1 \sin (\alpha_1-\theta_1)] \sin \theta_2$$

$$- [\rho_2 \sin (\alpha_2-\theta_2)] \sin \theta_1$$

and

$$y^* \sin (\theta_2-\theta_1) = [\rho_1 \sin (\alpha_1-\theta_1) \cos \theta_2$$

$$- [\rho_2 \sin (\alpha_2-\theta_2)] \cos \theta_1 .$$

Using the point determined by two lines of bearing as the initial estimate was suggested by a similar procedure described in Reference 3.

32

APPENDIX 2: Program Descriptions

PEST is a program that implements a position estimation procedure that is described in Appendix 1. With two or more observations on or from an object from or on two or more stations, the program will generate a position estimate for the object with an associated confidence region. After program initiation, the program user is presented with two options: (1) input bearings from stations on an object, (2) input bearings on stations from an object. To choose the first option, press T or t. To choose the second option, press S or s. Bearing information is entered in the following order: (1) the observed bearing on or from a station, (2) the bearing of the station from a reference location, (3) the range of the station from the reference location and (4) the bearing error. The reference location can be a station location. In this case, the station's bearing and range from the reference location are both taken to be zero. After two observed bearing observations and their associated station and bearing error information have been input, the user is given the option of (1) generating an estimate or (2) continuing to enter bearing observation data. A position estimate for an object is specified in terms of its range and bearing from the reference location. An associated elliptical confidence region which is centered on the estimate is specified in terms of the length and direction of its major axis, the length of the its minor axis and the probability (confidence) that it contains the object. The program user is given the option of defining the confidence region in terms of (1) size or (2) containment probability. For example, the size of an ellipse with a containment probability of .8647 is 2. In this case, the ellipse is referred to as a two sigma ellipse, since the semi-major and semi-minor axes of the ellipse are two standard deviation units in length. To choose the first option, press S or s. To choose the second option, press P or p. The estimate can be recalled by pressing E or e. Additional observations can be entered by pressing C or c. To quit, press Q or q.

The position estimation procedure requires an initial estimate of the object's position. In the program, the initial estimate is at the intersection of the two bearing lines that correspond to the first two bearings that are input to the program. Because of this, the first two bearing inputs should be from or on the two stations that are estimated to provide the best initial estimate. Note, if bearing errors are large relative to the angular separation of the two stations as seen from the object, the bearing lines from the two stations may not intersect. If this is the case, the reciprocal bearing lines will intersect and a gross error in the final position estimate could occur.

The procedure which is essentially equivalent to one described in Reference 2 is based on the following assumptions: (1) Bearings are taken on or from an object of unknown position from or on two or more stations of known position. (2) The distances involved are such that the object and the stations

35

can be considered to be located on a plane surface (a flat earth).
(3) The values of observed bearings on or from stations are
determined by independent normal random variables each with a
known standard deviation (the bearing error) and a mean equal
to the true bearing (zero bias). (If there is bias, it is known
and removed.)

COMP is a program that implements a procedure for combining a group of position estimates for an object. In order to use the procedure, the estimates must have associated elliptical confidence regions. When used, the procedure combines the estimates and their associated confidence regions into a single composite position estimate and associated elliptical confidence region. After program initiation, the program user is prompted to input the number of elliptical areas (confidence regions) to be combined. The user is then presented with two options for specifying the areas: (1) by containment probability or (2) by size. For the first option, the areas are specified by the probability (confidence) that they contain the object. This option is chosen by pressing P or p. For the second option, the areas are specified in terms of ellipse size. This option is chosen by pressing K or k. The size of an ellipse with a containment probability of .8647 is 2. In this case, the ellipse is referred to as a two sigma ellipse, since the semi-major and semi-minor axes of the ellipse are two standard deviation units in length. The orientation of an ellipse is specified in terms of the direction (the angle delta in the program) of its major axis. The position estimates, the centers of the elliptical areas, are specified in terms of latitude and longitude. The user is given the option of determining a confidence region in terms of (1) its containment probability (confidence) or (2) its size. To choose the first option, press P or p. To choose the

second option, press K or k. To recall the position estimate,
press E or e. To quit, press Q or q.

   The procedure which is described in detail in Section III is
based on the following assumptions: (1) The individual estimates
are determined by independent normally distributed random
vectors with known covariance matrices and a common but unknown
mean vector whose components are the object's rectangular
coordinates. (2) The object is located in the plane of the
coordinate axes, a plane tangent to a spherical earth. In
the program, the coordinates of the first entry are the coordinates of the point of tangency. Because of other uncertainties,
this flat search assumption should not introduce significant
estimation errors for the distance scale for which the program
is intended. In a sense, the procedure is a generalization of
the procedure described in Reference 4 which is limited to
circular confidence regions.

APPENDIX 3: Program Listings

```
10 CLS: SCREEN 0,0: DIM A(7)
20 PRINT "T=bearings on the target.": PRINT "S=bearings on the
stations."
25 A$=INKEY$
30 IF A$="T" OR A$="t" THEN JJ=0: GOTO 60
40 IF A$="S" OR A$="s" THEN JJ=1: GOTO 60 ELSE GOTO 25
60 CLS:INPUT "observed bearing";P: P=P*ATN(1)/45: IF JJ=1 THEN
P=P+180
70 INPUT "station bearing";Q: Q=Q*ATN(1)/45:INPUT "station
range";R: INPUT "bearing error";O: O=O*ATN(1)/45
80 IF I=2 THEN GOTO 140
90 I=I+1: A(I-1)=P: A(I+1)=Q: A(I+3)=R: A(I+5)=O: IF I=1 THEN
GOTO 60
100 X=A(4)*SIN(A(2)-A(0)): Y=A(5)*SIN(A(3)-A(1)): Z=SIN(A(1)-A(0)
): IF Z=0 THEN SOTO 330
110 U=(X*SIN( A(1))-Y*SIN (A(0)))/Z: V=(X*COS (A(1))-Y*COS( A(0))
)/Z
120 FOR M=0 TO 1
130 P=A(M): Q=A(M+2): R=A(M+4): O=A(M+6): GOSUB 390: NEXT M: SOTO
150
140 GOSUB 390
150 CLS: PRINT: PRINT "E=Est     C=Cont"
160 A$=INKEY$: IF A$="E" OR A$="e" THEN GOTO 180
170 IF A$="C" OR A$="c" THEN GOTO 60 ELSE GOTO 160
180 CLS: F=(B*B-A*C): IF F=0 THEN GOTO 380
190 X1=U+(B*E-C*D)/F: Y1=V+(A*E-B*D)/F: GOSUB 440: K=R1: J=B1
200 T=SGN(B)*ATN(1): IF A=C GOTO 220
210 T=.5*ATN(2*B/(A-C))
220 G=(C*COS(T)*COS(T)-2*B*COS(T)*SIN(T)+A*SIN(T)*SIN(T))/-F:
G=SQR(G)
230 H=(C*SIN(T)*SIN(T)+2*B*COS(T)*SIN(T)+A*COS(T)*COS(T))/-F:
H=SQR(H): IF H>=G GOTO 250
240 Z=H: H=G: G=Z: T=T+2*ATN(1)
250 CLS: PRINT: FRINT USING "\          \####.##";"bearing=",J*45/
ATN(1)
255 I1=K: GOSUB 500:   PRINT "range=";I1: PRINT
260 FRINT "S=Size P=Prob E=Est C=Cont Q=quit"
265 A$=INKEY$
270 IF A$="S" OR A$="s" THEN CLS: GOTO 320
280 IF A$="P" OR A$="p" THEN CLS: GOTO 350
290 IF A$="E" OR A$="e" THEN CLS: SOTO 250
300 IF A$="C" OR A$="c" THEN CLS: GOTO 60
310 IF A$="Q" OR A$="q" THEN END ELSE GOTO 265
320 INFUT "size";S: IF S<=0 THEN GOTO 320
330 O=1-EXP(-S*S/2)
340 FRINT USING "\            \##.####";"probability=",O: GOTO
370
350 INPUT "probability";O: IF O>=1 OR O<=0 THEN GOTO 350
360 S=SQR(-2*LOG(1-O)): PRINT USING "\       \####.##";"size=",S
370 X=S*G: I1=2*X: GOSUB 500:PRINT "major axis=";I1
371 N=T*45/ATN(1): IF N<0 THEN N=N+180
372 PRINT USING "\        \####.##";"direction=",N
375 Y=S*H:I1=2*Y: GOSUB 500: PRINT "minor axis=";I1: I1=4*ATN(1)*
X*Y: GOSUB 500: PRINT "area=";I1: GOTO 260
380 FRINT "no solution": END
```

```
X1=U-R*SIN(Q): Y1=V-R*COS(Q): GOSUB 440
W=P-B1: L=R1*O: IF L=O THEN GOTO 380
G=COS(B1)/L: H=SIN(B1)/L: IF W>=4*ATN(1) THEN W=W-8*ATN(1):
O 430
IF W<=-4*ATN(1) THEN W=W+8*ATN(1)
W=W/O: A=G*G+A: B=G*H+B: C=H*H+C: D=W*G+D: E=W*H+E: RETURN
R1=SQR(X1*X1+Y1*Y1): IF R1=O THEN B1=O: RETURN
IF ABS(X1/R1)=1 THEN M1=SGN(X1)*ATN(1)*2 ELSE M1=ATN(X1/R1/
(1-X1*X1/R1/R1))
IF ABS(Y1/R1)=1 THEN B1=2*ATN(1)*(1-SGN(Y1)) ELSE B1=2*ATN(1)
N(Y1/R1/SQR(1-Y1*Y1/R1/R1))
IF M1<0 THEN B1=8*ATN(1)-B1
RETURN
I1=100*I1: I1=INT(I1): I1=I1/100
RETURN
```

```
5 CLS: PI=4*ATN(1): INPUT "Number of Elliptical Areas";AO: PRINT
10 PRINT "Area Definitions:":PRINT
11 PRINT "By Containment Probability, Press P.": PRINT
12 PRINT "By Sigma Size, Press K."
15 E$=INKEY$
16 IF E$="P" OR E$="p" OR E$="k" OR E$="k" THEN GOTO 17 ELSE GOTO
   15
17 CLS: PRINT "For the Latitude and Longitude entry format, press
   F.  Otherwise, press C."
18 K$=INKEY$
19 IF K$="C" OR K$="c" THEN GOTO 26
20 IF K$="F" OR K$="f" THEN GOTO 23 ELSE GOTO 18
23 CLS: PRINT "Latitude and Longitude are entered in degrees and
   minutes and tenths of minutes in the form:  DDD-MM.MX  where X is
   N,S,W or E.  Leading zeros are optional, but the  -  must be
   included.  Press C to continue."
24 C$=INKEY$
25 IF C$="C" OR C$="c" THEN GOTO 26 ELSE GOTO 24
26 CLS:FOR O=1 TO AO
27 EE=0: PRINT
28 INPUT "LAT";A$: GOSUB 410: IF EE=1 THEN GOTO 27
29 Y=VA
30 INPUT "LONG";A$: GOSUB 410: IF EE=1 THEN GOTO 27
31 X=VA
32 IF O>1 THEN GOTO 34
33 YO=Y: XO=X: CO=COS(YO*PI/180)
34 Y=(Y-YO)*60: X=(XO-X)
35 IF X>270 THEN X=X-360
36 IF X<-270 THEN X=X+360
37 X=X*60*CO
38 INPUT "Delta";S
40 IF S>=180 OR S<0 THEN GOTO 35
45 INPUT "Major Axis";P: INPUT "Minor Axis";Q: P=P/2: Q=Q/2
50 IF E$="K" OR E$="k" THEN GOTO 60
55 INPUT "P";P3: R=SQR(-2*LOG(1-P3)): GOTO 70
60 INPUT "K";R
70 IF S>=135 THEN  S=S-180:T=P :P=Q: Q=T: GOTO 110
80 IF S>=45 THEN S=S-90: GOTO 110
90 IF S>=0 THEN T=P: P=Q: Q=T: GOTO 110
110 P=P/R: Q=Q/R
120 S=S*PI/180
130 U=P*P*COS(S)*COS(S)+Q*Q*SIN(S)*SIN(S)
140 V=P*P*SIN(S)*SIN(S)+Q*Q*COS(S)*COS(S)
150 W=(Q*Q-P*P)*SIN(S)*COS(S)
160 Z=U*V-W*W: A=A+V/Z: B=B-W/Z: C=C+U/Z: D=D+V/Z*X-W/Z*Y
170 E=E-W/Z*X+U/Z*Y: F=F+V/Z*V/Z*U: G=G+W/Z*W/Z*U
180 H=H-V/Z*W/Z*U: I=I+W/Z*W/Z*V: J=J+U/Z*U/Z*V
190 K=K+W/Z*U/Z*V: L=L-V/Z*W/Z*W
200 M=M+(V/Z*U/Z+W/Z*W/Z)*W: N=N-W/Z*U/Z*W
210 NEXT O
215 CLS
220 O=A*C-B*B: X=(C*D-B*E)/O: Y=(A*E-B*D)/O
230 R=(C*C*(F+I+2*L)-2*C*B*(H+K+M)+B*B*(G+J+2*N))/(O*O)
```

```
S=(A*A*(G+J+2*N)-2*A*B*(H+K+M)+B*B*(F+I+2*L))/(O*O)
T=((A*C+B*B)*(H+K+M)-C*B*(F+I+2*L)-B*A*(G+J+2*N))/(O*O)
Y=YO+Y/60: VA=Y: GOSUB 470: GOSUB 500
PRINT "LAT = ";L$
X=XO-X/(60*CO): VA=X: GOSUB 470: GOSUB 540
PRINT "LONG = ";G$
A=SGN(T)*ATN(1): IF S=R GOTO 280
A=.5*ATN(2*T/(S-R))
B=R*COS(A)*COS(A)-2*T*COS(A)*SIN(A)+S*SIN(A)*SIN(A)
C=R*SIN(A)*SIN(A)+2*T*SIN(A)*COS(A)+S*COS(A)*COS(A): A=A*180/

IF B>=C THEN D=A+90: GOTO 350
E=B: B=C: C=E: IF A<0 THEN D=A+180: GOTO 350
D=A: GOTO 350
PRINT: INPUT "P";P3: K=SQR(-2*LOG(1-P3)): F1=1: GOTO 330
PRINT: INPUT "K";K: P3=1-EXP(-K*K/2): F1=0
CLS: PRINT "Major Axis = ";2*K*SQR(R)
PRINT "Minor Axis = ";2*K*SQR(S)
PRINT "Delta = ";D: IF F1=1 THEN GOTO 345
PRINT "P = ";P3: GOTO 350
PRINT "K = ";K
PRINT: PRINT "E = Est   F = P   K = K   Q = Quit"
A$=INKEY$
IF A$="E" OR A$="e" THEN GOTO 390
IF A$="P" OR A$="p" THEN GOTO 320
IF A$="K" OR A$="k" THEN GOTO 325
IF A$="Q" OR A$="q" THEN GOTO 400 ELSE GOTO 355
CLS: PRINT "LAT = ";L$: PRINT "LONG = ";G$: GOTO 350
CLS: END
VA=VAL(A$)
D$=MID$(A$,4,1): IF D$="-" THEN GOTO 435
D$=MID$(A$,3,1): IF D$="-" THEN GOTO 430
D$=MID$(A$,2,1): IF D$="-" THEN GOTO 425
CLS: PRINT "Data entry error, restart entry for the area.":
1: RETURN
M$=MID$(A$,3): GOTO 440
M$=MID$(A$,4): GOTO 440
M$=MID$(A$,5): GOTO 440
VM=VAL(M$): VA=VA+VM/60
R$=RIGHT$(M$,1)
IF R$="S" OR R$="s" OR R$="E" OR R$="e" THEN VA=-VA
RETURN
QQ=SGN(VA): VA=ABS(VA)
IF VA>180 THEN VA=(VA-360) ELSE GOTO 480
QQ=-QQ: VA=ABS(VA)
DD=FIX(VA): FF=VA-DD: MM=-FIX(6000*FF)/100
RETURN
IF QQ=1 THEN Q$="N": GOTO 520
IF QQ=-1 THEN Q$="S" ELSE Q$=" "
L$=STR$(DD)+STR$(MM)+Q$
RETURN
IF QQ=1 THEN Q$="W": GOTO 560
IF QQ=-1 THEN Q$="E" ELSE Q$=" "
G$=STR$(DD)+STR$(MM)+Q$
RETURN
```

44

References

1. Bowditch, N., American Practical Navigator, Vol. I, Appendix Q, Defense Mapping Agency Hydrographic Center, 1977.

2. Daniels, H. E., "The Theory of Position Finding," J. Royal Stat. Soc. (B), Vol 13, 1951, pp. 186-207.

3. Thompson, K. P. and Kullback, J. H., "Position-Fixing and Position-Predicting Programs for the Hewlett-Packard Model 9830A Programmable Calculator," NRL Memorandum Report 3265, Naval Research Laboratory, Washington, D.C., April 1976.

4. "Navigation Lock (NAVLOC) Procedure (U)," Fleet Mission Program Library, Identification Number: U30089/A, Navy Tactical Support Activity, Box 1042, Silver Spring, MD 20910, 1 April 1984.

INITIAL DISTRIBUTION LIST

Copies

Director of Research Administration   1
Code 012
Naval Postgraduate School
Monterey, California 93943

Defense Technical Information Center   2
Cameron Station
Alexandria, Virginia 22314

Library, Code 0212   2
Naval Postgraduate School
Monterey, California 93943

Chief of Naval Operations   1
Attn: Code OP-953C2
Washington, DC 20350

Navy Tactical Support Activity   2
Attn: C. H. Earp, C. Reberkenny
P. O. Box 1042
Silver Spring, Maryland 20910

Commander   2
Submarine Development Squadron 12
Naval Submarine Base, New London
Groton, Connecticut 06340

Commander   2
Surface Warfare Development Group
Naval Amphibious Base, Little Creek
Norfolk, Virginia

Antisubmarine Warfare Systems Project Office   1
Department of the Navy
Washington, DC 20360

Naval Air Development Center   1
Johnsville, Pennsylvania 18974

Naval Surface Weapons Center   1
White Oak
Silver Spring, Maryland 20910

Naval Underwater Systems Center   1
Newport, Rhode Island 02840

Lightning Source UK Ltd.
Milton Keynes UK
UKHW012012021218
333216UK00014B/2500/P